The First Reading Books

The First Reading Books are fun reading books that fill the need for easy-to-read stories for the primary grades. The interest appeal of these true stories will encourage independent reading at the very beginning level.

The stories are told almost entirely with the 95 Commonest Nouns and the easier 110 words of the Dolch Basic Sight Words. Beyond these simple lists, the books use about one new word per page.

This Series has been prepared under the direction and supervision of Edward W. Dolch, Ph.D., University of Illinois.

PRINTED IN THE UNITED STATES OF AMERICA

Big, Bigger,

A First Reading Book

by EDWARD W. DOLCH

and MARGUERITE P. DOLCH

illustrated by Fran Matera

GARRARD PUBLISHING COMPANY
CHAMPAIGN, ILLINOIS

Biggest

Old Timer

Some deer are no bigger than little dogs. Some are much bigger than that. But the biggest deer of all is taller than a horse. The biggest deer is called a moose.

There was a man called Captain who was taking some men into the woods to hunt. These men had not hunted deer before.

Now all good hunters know that they must not shoot a mother moose. If the mother moose is killed, there is no one to look after the baby moose. But these men were not good hunters.

The men got excited when they saw a cow moose in the woods. They had never seen a deer that was so big. And before the Captain could stop them, the men fired their guns and killed the cow moose.

The Captain found the baby moose that was only about a day old. He took the little moose home to his farm. He put it into the barnyard with the horses and cows.

There was one cow who had lost her calf. She took the moose calf as her baby. She let it have her milk. She would not let any of the animals in the barn yard hurt her baby.

The Captain called the little moose Old Timer.

Old Timer Grows Up

Old Timer grew up on the farm. He liked all the animals that lived in the barnyard. He even liked the ducks and chickens.

Old Timer liked the Captain too. The Captain was always giving Old Timer something good to eat.

Old Timer grew and grew. Soon, he was taller than the cow that looked after him.

Then Old Timer found out that he could jump. Every day he jumped over the barnyard fence.

When Old Timer jumped over the barnyard fence, he did not run away into the woods. He only went to the kitchen door.

Old Timer knew that the Captain's wife would give him something good to eat.

One morning, Old Timer had jumped over the fence and gone to the kitchen door.

The Captain's wife went to the kitchen door with some very hot water. She did not know that the moose was at the door. And when she threw out the hot water, she threw it right on Old Timer. Old Timer was badly hurt on his shoulder.

The Captain put something on the hurt, and Old Timer was soon all right again. But when the hair grew over the hurt, the hair was white.

Old Timer was the only moose that ever had white hair on his shoulder.

Old Timer grew and grew. Soon the moose was taller than any of the horses in the barnyard. And now Old Timer wanted to go into the woods.

The Captain knew that before long, Old Timer would go into the woods.

One morning, Old Timer was not waiting at the door of the kitchen.

Old Timer was not in the barnyard. The Captain knew that Old Timer had gone into the woods.

Old Timer never came back to the farm. But sometimes when the Captain was hunting in the woods, he would see a big moose. If the Captain was very still, the big moose would come nearer. And the Captain could see that the big moose had some white hair on his shoulder.

Queen

The elephant is the biggest animal of all. Queen was not the biggest elephant in the circus. But Queen was a very big elephant.

The circus wild animals were in cages that were on wagons. Each wagon must be put in its place. And Queen knew just where to put each wagon.

There was one wagon that Queen would not go near. It was a big wagon. In the big wagon was a rhinoceros. A rhinoceros is a very big and dangerous animal. A rhinoceros will fight any animal, even a big elephant. Queen would not go near that rhinoceros.

In the circus, Queen and a clown always made the people laugh.

The clown would take Queen's trunk and talk into it just as if he were talking into a telephone.

As soon as the clown had "talked into the telephone," Queen would turn around. Then the clown would take Queen's tail and hold it to his ear just as if he were using a telephone.

The clown would laugh and laugh just as if he were hearing something very funny over the telephone.

The people at the circus always laughed at Queen and the clown.

After many years, the clown, who was getting old, left the circus. He lived in New York. But if a circus came to New York, the clown always went to see it.

One day the clown went to a circus. He was not dressed as a clown now. He was standing near the elephants talking to some people.

Someone was telling the clown a funny story. The clown was laughing and laughing.

All at once, one of the elephants made a great noise. The men could not get the elephant to be still. The clown went over to the elephant to see what was going on.

It was Queen who was making all the noise. She knew it was her clown. And the clown knew what to do.

The clown took Queen's trunk and talked into it just as if it were a telephone. Then the big elephant turned around. The clown took the elephant's tail and put it to his ear just as if he were using a telephone.

Queen was happy to see her clown again. But the funny thing was that the elephant had not seen her clown for twenty years.

The elephant heard the clown laugh and she remembered what they had done twenty years before.

The Bears at Yellowstone

The grizzly bear is the biggest of all bears. He lives in the Rocky Mountains. At Yellowstone Park, in the Rocky Mountains, no one can shoot an animal.

So the black bears and the grizzly bears know that no one in the park can hurt them.

And the bears know that there is one place where they can get good things to eat.

Every day the bears come to where the garbage from the Park Hotel is dumped.

The black bears stand around and wait for the man to come with the garbage cans. People go to the garbage dump to take pictures of the bears.

When the grizzly bears come to the garbage dump, the black bears do not go near them. Every animal that lives in the mountains knows that the grizzly bear is the fiercest of them all.

There was one grizzly bear that lived in Yellowstone Park that was not afraid of anything. He would walk into the kitchen of the hotel to get something to eat. The men had to catch this bear with a rope. They sent him to the zoo, for they were afraid that this grizzly would hurt someone.

The Bear and the Can

One day a grizzly bear came to the garbage dump. He found some very good things to eat. Then he found a can with something very good in it.

The bear put his paw into the can to get more of the good things. And then the bear found that he could not get his paw out of the can.

The grizzly went away from the dump. But the men at the Park could hear him. Clump, clump, clump, went the can on his paw.

The big bear growled and growled all the time. He wanted to fight anything that got in his way. All the other bears got out of his way.

For the bear's paw hurt him very much. He could not get his paw out of the can.

The men at the Park did not know what to do. They heard the bear going clump, clump, clump. They heard him growl. They wanted in some way to get that can off the bear's paw. But to do this, they would have to catch the bear first. And how could they do that?

Then some cowboys came to the Park, and the men knew what to do. The two cowboys had their ropes and they knew how to use them.

The two cowboys went to the garbage dump and waited for the grizzly.

The grizzly came to the garbage dump and soon was eating. Then the two cowboys threw their ropes. They caught the grizzly with two ropes. Then they tied the grizzly up so that he could not hurt anyone.

The Park men came and took the can off the paw of the big grizzly.

Oh, how the big grizzly growled, for his paw hurt him very much. The grizzly wanted to kill the men.

The cowboys had to be very careful how they let the grizzly go. He was going to fight. But the cowboys jumped on their horses and away they went.

The big grizzly went into the woods and stayed there until his paw was all right. And he never put his paw into a can again.

A Strange Animal

The giraffe is the tallest of all the animals. It is sometimes twenty-three feet from its feet to the top of its head. It is a very strange-looking animal. It has long legs and a very, very long neck.

The very, very long neck of the giraffe lets it get what it likes to eat. For the giraffe eats the leaves from the tops of trees.

The giraffe has no way of making any sound. The giraffes stay together. When one wants to tell that danger is near, he makes his tail go very fast. He can talk only with his tail.

Giraffes do not want to fight. They do not even fight each other. When danger is near, they try to run away. The only way they can fight is by kicking out with their feet.

A mother giraffe has a hard time looking after her baby. The baby giraffe wants to find out about everything. And the mother has a hard time making the baby know that there is danger about.

When the baby goes away, the mother giraffe cannot call it back. For the giraffe cannot make a sound. The mother has to go after her baby. Then she walks around her baby and pushes it away from the danger.

The lion is the greatest danger to the giraffe. The lion likes to eat a baby giraffe. But the giraffe has very good eyes. And it is always looking for a lion.

A funny story is told about this strange animal.

At one time, there was a giraffe in a zoo. His head was way up on his very, very long neck. His legs were very long.

A man who had never gone to a zoo before was looking at the giraffe. This was the first time he had ever seen a giraffe.

The man looked and looked. He looked at the giraffe up and down.

"You can't fool me," the man said. "There is no animal that looks like that."

Bobo, a Baby Gorilla

The gorilla is the biggest of all monkeys. You sometimes see a gorilla in a zoo or in a circus. But the gorilla is always in a very strong cage.

Gorillas live in Africa. They are very hard to catch.

The old gorilla is taller than a man. He weighs more than three times as much as a man.

Gorillas live in families. And they are very fierce.

Mr. and Mrs. Hoyt were in Africa hunting wild animals for a museum. They also were taking pictures of wild animals. Mr. Hoyt wanted to get a big gorilla for the museum. And he wanted some pictures of live gorillas.

One day, the natives found a family of gorillas. The natives came to Mr. Hoyt.

They said that they would take Mr. Hoyt to the gorillas. So Mr. and Mrs. Hoyt went with the natives.

The natives of Africa like to eat the meat of gorillas. Mr. Hoyt said that he would shoot a big gorilla for them. He wanted to get the skin for the museum. The natives wanted the meat.

When they got to the place where the family of gorillas lived, they found many natives.

It was hard to keep the natives still so that pictures of the gorilla family could be made. Then Mr. Hoyt killed a very big gorilla. It was the biggest one that he had ever seen.

The natives were so excited that they killed all of the big gorillas. But one mother gorilla had a little baby. So the natives took the baby gorilla to Mrs. Hoyt.

Mrs. Hoyt looked after the little baby gorilla. She had a native boy stay with the baby gorilla all the time so that nothing would hurt it. The little gorilla had its milk like a baby. It wore clothes like a baby. The baby gorilla was called Bobo.

When Mrs. Hoyt left Africa she took Bobo with her. And Mrs. Hoyt took the native boy who had looked after Bobo.

Bobo Grows Up

Bobo, the baby gorilla, grew up like any baby. She had her bedroom. She learned to walk.

Bobo wore little dresses. She wore overalls to play in. But when Bobo got hot, she would take off all her clothes.

Bobo learned to hold Mrs. Hoyt's hand when they went for a walk in the garden.

Bobo did not learn to read, but she liked to look at pictures in a book.

Bobo learned to use a chair and to eat at a table. She liked to eat just what any little girl eats. She learned to ride a tricycle.

Mr. Hoyt made a playroom in the garden for Bobo. She liked to play just like a little girl does. She liked music, and she learned to dance in time to the music.

But Bobo was not always a good little girl. When Bobo was three years old, she was as strong as a big man.

Someone had to be with Bobo all the time. When she was not happy, she would break things.

Bobo had one place in the garden where she liked to go and be in the sun. One day Bobo went to this place and found some kittens there.

Bobo looked at the kittens a long time. She put her hand on each little kitten. But she was careful not to hurt the kittens. At last she took up a little white and black kitten and took it to the house.

Bobo took the kitten to her bedroom. When Bobo went outdoors to play, she carried the kitten with her. The kitten grew to be a cat. And Bobo still carried the cat around the house.

One day, the cat had kittens in Bobo's bedroom. Bobo must have known that the mother cat liked the kittens more than she liked Bobo. For Bobo never played with that cat any more.

But Bobo took one of the black and white kittens. She carried this kitten around all the time. And when the kitten grew to be a cat, she never left Bobo.

Bobo grew bigger and stronger all of the time. She sometimes threw tables and chairs about. A push from Bobo would hurt one very badly.

One day just in fun, Bobo pushed Mrs. Hoyt. Mrs. Hoyt was badly hurt. And Mr. Hoyt said that Bobo was too big and strong to live with them.

Bobo must be sent to a zoo where she could be put in a strong cage.

At last, Mrs. Hoyt let the Ringling Brothers Circus have Bobo as a wife for their big gorilla, called Gargantua.

Catching a Wild Animal

You must remember that the wild animals that you see in cages at the circus or at the zoo had to be caught.

Mr. Mayer was a hunter who caught wild animals for the zoo or for the circus. He went to Malaya where there are many wild animals.

There is one big animal that you sometimes see in a zoo or in a circus.

This animal is always in a very strong cage. This big animal is a rhinoceros, sometimes called a rhino.

The rhinoceros is a very dangerous animal. He lives where the elephant lives. And even the elephant does not want to fight a rhino.

The natives of Malaya helped Mr. Mayer. He told them how to make a pit-trap to catch wild animals.

When an animal gets into a pit-trap, he is not hurt.

Nets are also used to catch wild animals. The natives knew how to make nets out of a vine that is very strong. And the natives knew how to make strong cages out of small trees tied together with these vines.

Mr. Mayer had caught many wild animals in Malaya. Now he wanted to catch a rhino.

It is very dangerous to try to get a live rhino, and to get him into a cage. For the rhinoceros is very big and very strong.

One day, some natives came out of the woods. They told Mr. Mayer that a rhinoceros was in the woods. Mr. Mayer and the natives went into the woods. Mr. Mayer showed the natives where to make the pit-traps. He showed them where to put the nets.

Then a native with a drum had to watch each pit and each net. The natives were up in trees. When any animal got into a pit or was caught in a net, a native hit his drum. Then Mr. Mayer and the natives came.

Mr. Mayer waited a long time. Then one morning he heard the drums. He knew that some animal was caught in a net. And he went to see if the animal was a rhinoceros.

A Big Mother

When Mr. Mayer and the native helpers got near to the net, they heard a great noise. They knew that the noise was being made by a rhinoceros.

One of the natives went on very carefully. He did not want the rhinoceros to smell him. The native in the tree called out. He said that a baby rhino was caught in the net.

The big mother was going around and around the net trying hard to break it.

But the mother rhinoceros smelled the native who was in the grass. She turned and ran at him. He had just time to get up into a tree.

Now the rhinoceros could smell men all around her. She ran back to her baby who was caught in the net.

Mr. Mayer told two natives to go and get two more nets. He wanted to catch the mother rhino as well as the baby.

But he must try to keep the mother from going away from the baby who was caught in the net. And so he had the man in the tree by the net cut a long stick. When the mother wanted to go off into the woods, to hunt the man smell, the man hit the baby rhino with the stick and made it cry.

Then the mother went right back to the baby.

The natives came back with the two nets. Mr. Mayer put the two nets together, for it would take a very strong net to hold that mother.

The nets were put up so that when the rhino ran at the men, she would be caught in the nets.

At last everything was ready. The men were by the nets.

Mr. Mayer was ready with his gun if the nets did not hold the rhino.

Mr. Mayer told the men to be very still. The baby rhino was so tired that it was still too. The woods were very still. The big mother could not see very well, but she could smell men.

Then one of the men by the nets called out. The mother turned and ran at the man.

The rhino was going to kill that man. But the net was in the way. The big mother was caught in the net.

The natives were so happy that they called out and danced. But there was much work to be done. Mr. Mayer put the natives to work to make a strong cage to hold the big mother and her big baby. In this cage, they were sent back to America.